I0490584

SOLAR
INSTALLATION AND
OPERATIONAL BOOK

Practical guide on solar thermal and photovoltaic panels

Austen C. Hamilton

Table of Contents

CHAPTER ONE

INTRODUCTION FOR SOLAR

Solar component parts usage. Solar thermal electrical energy is normally used for water heating. It's a convenient technology; the panels on your roof are the collectors of sunlight, for that reason heating up the liquid in the tubes which is then transported into your cylinder geared up for use. Solar thermal constructions use the sun's power to warmness your water.

The collector plates on your roof take in the suns electrical energy and they swap this into each glycol or at as soon as into the water from your tank. It's then pumped alongside pipes to your heat water cylinder, heating up the water to a preset temperature.

CHAPTER TWO

PROPER INSTALLATIONS GUIDE

How To Solar Thermal System

Make a ladder to use while climbing up

You have to get on the roof. Trace out the format of the racks, Loosen the shingles the area the base plates will go, reduce the single away from the vicinity the base plate will sit. The base plate holds the rack that holds the panel. Put Caulk on the holes that the base plate will be drilled into.

Create a base plate

Screw on the base plate. Put flashing over the plate and beneath the shingles. This stops water from coming in. This is moreover why we loosened up the singles in. Bolt the L clamp to the base plate. The rail will be related to the L clamp.

Bolts rails and tighten the square

Repeat the process until you have all the L feet set up on all rows that you mapped out in. Put in rails loosely. Do now not tighten them until they're square. Bolts rails to L clamps and insert bottom toes that will hold the bottom lip of the panels. Bring up panels on elevate latter, put panel on rails and attach.

Install panel and thermal censor

Install thermal censor and intakes outputs from the panel through the roof, the line-set the intakes and outtakes of the panels is then snaked to the current heat water machine generally in the basement.

Connect the pump

The plumber will interconnect the system, stress takes a appear at it, fill it with glycol and then you'll flip on the monitoring machine that will taking walks the pump and the desktop is operational. Almost every section of the photograph voltaic thermal set up is fundamental to usual overall performance and durability. L-Foot Attachment, Each L foot ought to be firmly hooked up to rafters for structural support, specially wind and snow loading.

Ensure the Square Rail is place well

For aesthetic features its key that the rail is square. Also, many photograph voltaic thermal producers pre-measure each rail so they can additionally no longer go well with flawlessly with the panels you've different if it's now no longer square. It's in all probability that if the rail is now no longer rectangular the panels will no longer fit.

Check the connect system for proper flow of fluid

The connection between panels is ought to for pressurized systems. If the connection is no longer 100 percentages exceptional fluid will leak from the laptop and it will begin to feature poorly in a quick quantity of time. Good plumbers will pressure test the machine with air in the past than they fill it with glycol

Check the Penetration circle

The linset penetration is the thing the region most leaks will take region so it's necessary that the penetrations by using the roof and the membrane used will retain water out. Never used caulk as the predominant barrier between the out of doors and conditioned space.

Connect solar inverter to the panel

Solar inverters are the operational talent of photovoltaic (PV) systems, making them one of the most critical elements of a photograph voltaic system. Since image voltaic panels generate power in DC, which is now not really helpful for most home appliances, you will generally choose an image voltaic inverter.

Converts DC and AC in output

String inverters, moreover diagnosed as central inverters, are the oldest and most conventional type of picture voltaic inverter used today. They work via ability of connecting a string of image voltaic panels to one single inverter, which converts the entire DC enter into AC output. Because string inverters are the oldest kind of photograph voltaic inverters, they are moreover the most reliable.

Mount and cross check

After a lengthy time of being on the market, string inverters have had most of the kinks labored out. They are moreover the least excessive priced photo voltaic inverter option. String inverters are moreover centrally located on the factor of your dwelling or shut to the component of a ground-mount. This approves easier get admission to monitor, repair, or exchange the inverter.

Benefit of string inverters

While string inverters are reliable, they are moreover a whole lot much less surroundings pleasant at optimizing image voltaic electrical energy output.

Because string inverters are associated to a total string of photograph voltaic panels, shading on one photograph voltaic panel will minimize the electricity output of the total string. Also, string inverters fully furnish total-system monitoring as adversarial to panel-level monitoring. This can be a draw back when diagnosing issues with image voltaic production, and it can moreover be unfortunate for these photograph voltaic house owners who decide on a higher granular stage of monitoring.

Types of inverters

The Central inverter

Central inverters are same to string inverters then again they are lots massive and can assist greater strings of panels. Instead of strings going for walks at as soon as to the inverter, as with string models, the strings are linked jointly in a accepted combiner area that runs the DC electrical energy to the central inverter the location it is changed to AC power.

Central inverters require fewer element connections; alternatively require a pad and combiner box. They are first-rate fabulous for big installations with consistent manufacturing all through the array.

The Micro inverter

Micro inverters are moreover turning into a well-known want for residential and enterprise installations. Like energy optimizers, micro inverters are module-level electronics so one is set up on each panel. However, now not like electricity optimizers which do no conversion, micro inverters convert DC power to AC suitable at the panel and so don't require a string inverter.

Also, due to the truth of the panel-level conversion, if one or larger panels are shaded or are performing on a minimize stage than the others; the ordinary overall performance of the closing panels won't be jeopardized. Micro inverters moreover expose the usual overall performance of each personality panel, at the same time as string inverters showcase the common overall performance of each string. This makes micro inverters ideal for installations with shading troubles or with panels on a couple of planes dealing with a range of directions. Systems with micro inverters can be higher efficient, then again these often price greater than string inverters.

CHAPTER THREE

PHOTOVOLTAIC AND HOW TO INSTALL IT

Installations Of Photovoltaic Solar

Select a place to mount solar

First, you have to survey your property to pick out the fantastic place to set up the panels. Remember, genuinely due to the reality the photo voltaic hits and vicinity of your rooftop or property doesn't mean it's the fine spot. The pitch and direction of the roof have an impact on the of the picture voltaic modules.

Create equipment platform

Therefore, make advantageous you pick out neighborhood that will furnish the panels most publicity to the photo voltaic for most of the day. Step two entails making equipped the place for the placing of the modules. You can bring together the platform the use of metal or aluminum rails.

Have a measurement for mounting space

Design the mounting laptop with the measurement and width of the panels in mind. Make sure it is sturdy adequate so that it can face up to extreme weather. Building a platform requires that you run conduits from the thing of set up to indoors the house, the location the generated electrical energy will be processed.

Set the panels at locations chosen

The panels are set up on the platform the utilization of brackets, bolts, or clamps. The role of the add-ons is to make positive that the modules are held firmly on the platform. Be cautious when inserting in panels on the roof to preserve away from accidents and damages.

Connect the electrical part to the panel

You will prefer to wire the image voltaic modules at the same time the use of junction connectors or a fuse combiner box. Make positive that the cables are well insulated to preserve away from strength leakage and accidents.

Once complete, be part of the exterior wiring with the indoors manipulate panels. The connection will allow electrical energy to go with the drift from disconnect to a value controller and battery economic organization for storage.

Connect the inverters

Alternatively, it can waft from disconnect to an inverter. Whether electricity will first go with the waft into the fee controller or the inverter will depend on the form of picture voltaic electrical energy machine in place.

Reason for connecting earthing to solar panels

(1)Protect you and others from electric powered shock, which can be fatal

(2)Offer security in opposition to furnace that can be introduced on via the use of the system

(3)Protect picture voltaic aspect in opposition to lightning given that it motives an electrical energy surge.

Connect the electrical energy

You can then direct the electrical energy from the inverter to the crucial electrical control panel of the house. Everything ought to be appeared over as quickly as increased beforehand than initializing a test run. The rationale of the take a look at run is to make certain that the total aspect is working correctly. Once complete, it's time to shut and commence reaping the benefits of photovoltaic power.

What is PV Cell?

A PV Cell or Solar Cell is the smallest and foremost setting up block of a Photovoltaic System Solar Module and a Solar Panel. These cells vary in dimension ranging from about 1.5 inches to 4 inches. These are made up of photograph voltaic photovoltaic cloth that converts photograph voltaic radiation into direct cutting-edge (DC) electricity.

Types of PV Cells

The Crystalline Silicon PV Cells (Monocrystalline)

These Solar Cells are manufactured from crystalline silicon. Many of you ought to be grasping that silicon is the second most prevalent cloth on Earth and is abundantly positioned in sand. To make image voltaic cells out of silicon, manufactured silicon crystals are sliced to about 300 micrometers thick and protected to work as a semiconductor to capture image voltaic energy.

The Thin-film

Thin-film PV cells use amorphous silicon or a desire to silicon as a semiconductor. These photograph voltaic cells are enormously flexible and can be except lengthen set up with establishing materials. They work exquisite even in the path of clouds when there is low photo voltaic light.

Circle Conversion Of DC To AC Electricity

The PV cells generate DC or direct current. This DC electrical power has to be changed to AC or alternating present day so that it can be used in a home lights fixtures desktop or taking walks appliances. An inverter is used to convert DC to AC. This is equal as altering DC from a battery to AC.

Easy ways of storing electricity with photo voltaic cells

The electrical strength generated with the resource of picture voltaic cells by the utilization of image voltaic electrical energy has to be saved so that it can be used later as a when required. This is performed with the taking walks the cutting-edge into an economic organization of Solar Batteries. A single photovoltaic Module can produce very little electricity. This strength is too tons much less for use in any household or for an industrial purpose. Hence, an array of such PV Modules is electrically associated together to structure a massive Solar Photovoltaic Panel. A PV Panel can have any vary of PV Modules relying requirement of Solar Energy.

What is Solar Mount Structures?

Solar Mounting Structures are critical for the surroundings pleasant working of a picture voltaic energy gadget in every utility and rooftop applications. While the most balance of device

(BOS) components, alongside with inverters, DC cables, junction boxes, transformers, and so on.

Types of Solar Mount

Mounting On-roof

By a lengthy way the most customary type of image voltaic panel mounting is an on-roof system. As the title suggests, the photograph voltaic panels restore at as soon as to the roof. On-roof photograph voltaic panels are a low-cost solution.

Provide superb air go with the flow to your panels and gold widespread performance. When placing in an on-roof picture voltaic system, metal hooks are drilled right away into the rafters of the roof. This entails each now and then casting off and altering some roof tiles. Vertical rails run for the duration of the hooks, to which your Solar PV panels are clamped onto. Finally, a local weather proof seal is utilized spherical the hooks for most protection.

Mounting on Ballast

Mounts are same to flush mounts; besides they use weights to hold the picture voltaic panels on your roof in place. This format can maintain labor, time, and, money, on the other hand it moreover gives the problem of transferring the weights onto the roof, which can be notably difficult with giant systems.

Ballasted mounts take away roof penetrations, are quicker and lots much less high-priced to install, and allow for a panel tilt of up to 20 tiers for the notable photo voltaic exposure. This mount, on the unique hand, presents to the weight on your roof, has a minimize electricity density, and is a lot much less best for high-wind environments. Certain net website online characteristics, such as roof slope and establishing height, restrict them.

Mounting In-roof

In roof photo voltaic or built-in image voltaic panels are the first-rate reply for new builds or every person looking out to re-roof their home. Many purchasers determine for an in-roof gadget due to the truth of the sleeker aesthetics.

As the picture voltaic panel sits down snags internal a tray, there is no place for birds to nest under and the panel's exhibit up flush with the leisure of the roof. However, this does give up end result in a good deal much less air go with the flow spherical the panels, ensuing in 8% reduce efficiency.

Mounting Hybrid

Mounts are an aggregate of flush mounting and ballasted mounts, as you can also expect. Some structural parts from every mounts are used to accommodate roofs that can't help either.

Hybrid mounts choose little roof penetration, can be hooked up quickly depending on the model, and allow for specific design optimization chiefly primarily based on load-bearing and wind. These mount are typically greater excessive priced and take up greater residence on your roof, reducing the volume of location available for your system.

Mounting on tiles

Solar tiles are regularly seen the picture voltaic panel of the future. Unlike a common photograph voltaic panel which is hooked up on-top of your roof tiles, picture voltaic tiles trade your roof tiles all together. Seamlessly integrating image voltaic science into property design, Although Solar tiles may additionally show up the part, their efficiencies aren't especially on par with on-roof photograph voltaic panel machine certainly yet.

Mounting on Flat trapezoidal roofs

Suitable for flat trapezoidal roof structures, these metal triangular frames make positive that your picture voltaic panels are tightly secure.

Screwed to the flat roof in the equal way as a pitched trapezoidal roof, these triangular frames allow you to mind-set your photograph voltaic PV panels for fantastic performance. With a viable tilt viewpoint from 5° to 45°.